How to Build a Backyard
WILDLIFE POND

by
Theresa Berrie
at
Our Tiny Homestead

How to Build a Backyard Wildlife Pond
by Theresa Berrie

Text, Design & Layout by Theresa Berrie
Photography by Theresa & Rob (Bear) Berrie

© 2019 by Theresa Berrie
OurTinyHomestead.com
All Rights Reserved

Other books from Our Tiny Homestead:
*Eat Your Wild Yard-Spring:
Growing & Foraging for Wild Edibles in Your Own Backyard*

Thanks to my partner Bear,
for helping me follow my dreams
of turning our yard into gardens, even when he didn't want to.
Many years later, Mother Nature has won him over,
and I love seeing him get excited by every new wild creature
we discover in the yard. He even joined an
online group for spider lovers.

Table of Contents

Introduction: Why I Built a Wildlife Pond • 2

Pond Building Steps:
 1: Choose & Prepare Your Location • 4
 2: Create a Pond Design • 6
 3: Rough Dig the Pond • 8
 4: Level the Edges • 10
 5: Smooth the Bottom and Sides • 10
 6: Install Padding to Protect the Liner • 12
 7: Buy the Liner • 13
 8: Install the Liner • 14
 9: Fill the Pond with Soil, Rocks, and Water • 16
 10: Building a Privacy Trellis Behind the Pond • 18
 11: Adding a Frog Condo • 19
 12: Adding Rocks Around the Edge of the Pond • 20
 13: Our Finished Pond • 22

Extras:
 Additional Ways to Build Wildlife Habitat • 24
 Wildlife Seen at Our Pond • 26
 Dragonflies • 26
 Frogs • 28
 Pond Plants • 30
 Food from Our Pond: Arrowhead Tubers • 32
 Roasting Arrowhead "French Fries" • 34
 Natural Pond Maintenance • 36
 Winter Pond Maintenance • 40
 How to Patch a Hole in the Pond Liner • 42
 The Pond in Later Years & Through the Seasons • 44

Conclusion, To Learn More • 46

These baby frogs in our pond are about a centimeter long.

Why I Built a Wildlife Pond

When Bear & I moved into our house in 2005, it was a fixer-upper. Inside renovations took up most of our weekends, but I also tried to find time to work outside, since I fondly remembered gardening as a child. It has taken almost 15 years, but we have finally removed the lawn from our tiny yard (60'x140'), and replaced it with gardens that benefit both us and wildlife. Using permaculture principles that advocate sustainable living, we now have a landscape of perennial edibles that works in harmony with Nature, so also provides an abundance of wildlife habitat (see our website, *OurTinyHomestead.com,* for more).

Several years after moving in, **I certified our yard as wildlife habitat** with the National Wildlife Federation, because it provides sources of food, water, shelter, and places for wildlife to raise young. To qualify at that time, I needed either a natural water feature, such as a stream, or a bird bath (now other options count). I had already been feeding the birds, so including a bird bath was easy enough. Eventually, though, I got tired of filling it every day and it was difficult to keep clean. Also, I really wanted to fall asleep while listening to frog calls. So **I began looking for information about how to build a backyard wildlife pond.** I found lots of books about traditional ponds, but only a couple sources that advocated building a natural pond that would be more beneficial to wildlife (with soil in the bottom and no pump, see next page), and those books only had a couple pages of instructions. **I wanted others to have an easier time figuring out how to build this type of pond,** so I shared my experiences on our website, and got a lot of positive feedback. I have created this book to encourage even more people to join me in restoring a bit of habitat for the other species with whom we share our planet.

I've included lots of dos and don'ts that may seem complicated, to help you build a pond that will attract the most wildlife, while being as easy maintenance as possible. Build whatever you can, though, even if it doesn't follow all the rules (which is what I had to do). **Even if all you can do is dig a small hole, cover it with pond liner, and fill it with water, do it and see what happens.** It may need extra maintenance or may not last as long as a more complicated pond, but it will provide important habitat and give you the joy of interacting with more wildlife.

Building our pond was not very hard or very expensive. It cost about $300, and I did most of the work by myself during a week-long Spring vacation in 2012. My partner Bear sometimes helped when he was home from work, and was essential for the heavy lifting (of the rocks and pond liner). The result has been one of our most rewarding garden projects. **It's one of the best things we've done to create habitat in our yard, as water is the one essential element that benefits all wildlife.** It's also fascinating for us humans. It is an entrance to another world where we can't live, making it the most mysterious part of our yard. I am drawn to it more than to any other garden feature. **Having a pond outside our door lets me observe and interact with wildlife I wouldn't otherwise see. I hope you join me on this adventure and discover the joys of having your own backyard wildlife pond.**

~ Theresa Berrie, 2019

A wildlife pond is different from other backyard ponds:

- **It doesn't have a pump:** Pumps can kill small wildlife, such as frog tadpoles, when the pond water gets sucked through the pump. If you're familiar with any ponds that have pumps, you'll realize that regardless of this, they support frogs. But that doesn't necessarily mean the pump isn't killing tadpoles. It may mean that fewer of them survive. The water disturbance that pumps create is also hard on other wildlife, such as dragonflies, that prefer to breed in still water. Most people think a pond needs a pump so that the water doesn't become stagnant and smelly, and won't breed mosquitoes, however, circulating the water through a pump is not the only way to deal with these issues (see the section on *Pond Maintenance*).

- **It has soil in the bottom of the pond:** Most ponds constructed by humans have only rocks or pea gravel on the bottom and the plants are grown in pots. The pots are removed in the Winter and stored indoors, and the plants are repotted into larger pots each Spring. This was way more work than I wanted. Fortunately, in a wildlife pond you add soil to the bottom, in order to create better habitat for animals, and in which you can grow plants.

- **It doesn't have exotic goldfish:** Fish will eat frog eggs, and I wanted frogs in our garden because they help keep the ecosystem balanced by eating insects, beetles, slugs, and snails.

- **It's cheaper:** Pond pumps are expensive. Since we didn't need to buy one, our pond cost around $300, most of which was for the rubber liner. I bought an expensive liner because I wanted it to last as long as possible, but there are cheaper options available (see *Step 7*). Without a pump, our pond also doesn't increase our electric bill during the Summer, although you may need an electric pond heater in the Winter, depending on where you live.

- **It's easier to maintain:** Once I found pond plants that survived our Winters, our pond achieved a balanced state that requires little maintenance. Other than adding water when it hasn't rained for a while, I do very little.

A tadpole has turned into a tiny frog and is emerging from our pond.

Build it Even if Experts Tell You It Can't Be Done:

If you're reading about ponds or talking to staff in pond shops, you probably won't find much support for constructing a pond this way. In fact, like me, you might be told that it's impossible to maintain a pond without a pump or with soil in the bottom. However, it has worked in our yard. We've had our pond for seven years now, and the water stays clean, it doesn't breed mosquitoes, and we get to enjoy frogs and other wildlife as neighbors.

1. Pond Building Steps: Choose & Prepare Your Location:

Things to consider when deciding where to put your pond:

- **Most pond plants need full Sun** (at least 6 hours a day), so put your pond in a sunny area if possible.
- **Check the location of your underground utilities** so you won't damage them or yourself (in the case of electrical wires) while you're digging. In Wisconsin, we can contact the Digger's Hotline (*diggershotline.com*). They will mark the location of underground gas & electrical lines. There turned out to be none in our yard, so we realized we had to contact our Village office instead to find out where the water & sewer lines were. However, the records were so old they gave us the wrong information, and we almost put the pond directly over where we later found the water main. So the lesson I learned is to pay attention while you dig, no matter what.
- **Is it near an electrical outlet**, if you need a pond heater (see the *Winter Pond Maintenance* section)?

Try not to put your pond:

- **Where it will get herbicide or pesticide run off** such as downhill of your neighbors' yards.
- **Under a tree**, because you might run into big roots while digging the hole, and the water will get full of leaves in the Fall.
- **Near bird feeders**, so that bird seed doesn't fall into the water (ours gets a lot of bird seed blown into it when there's a strong wind).
- **Near walnut trees**. Walnut trees drip a substance that is meant to keep nearby plants from growing and competing with the tree. One book I read said it would kill everything in your pond. However, we have not had problems with this. Our pond is outside the farthest reach of our walnut tree's limbs, but it is close enough that it gets nuts and leaves from the tree in it in the Fall. It hasn't had any effect on the pond that I've noticed, but if you have the option, locate your pond farther away from walnut trees than we were able to.
- **Don't put it right next to your foundation**. It's just wise to keep water away from the basement in case there's a leak or the pond overflows.

☙ IMPORTANT ❧
Locate your pond where you'll easily see it so you can enjoy the wildlife it will attract:

- You'll want to have **space for a bench or other seating at the edge of the pond**, so you can look down into the water and see the small bugs, snails, and other wildlife that will be moving about. We didn't have a lot of space, so used tree stumps as not-entirely-comfortable seating.
- If you can **put it next to your deck or patio**, you'll be able to sit a bit farther away and maybe see larger visiting wildlife.
- If you can **locate the pond where you can view it from inside your house**, you'll get even more enjoyment out of it. Animals will be more likely to come to the pond when you're not outside and you can watch the pond during the Winter months too, when it will still be visited by birds and animals looking for a drink.
- The above might seem obvious, but I learned they weren't always easy to do. When we returned from a trip to Ireland with photos of a great backyard pond at one of our B&Bs, it inspired my mother-in-law Marge to build a large pond in her own yard. However, she wasn't able to view it from anywhere in the house, and even the railing on her back deck was at the exact height to block the view. You could hear the frogs from any open window, but it was always frustrating that there was no place to comfortably sit and enjoy watching the pond.

Our Pond Location:

We decided to put our pond in our side yard, which is about 20 feet wide (see the photo to the right). We chose the location directly across from the dining nook we built in our bay window, so we can watch the pond while we're sitting at the table. There were several disadvantages to this location (it was next to our bird feeders and near walnut trees), but there really wasn't any other place to put the pond in our yard that would let us enjoy it from inside the house.

To prepare your chosen location:

1. Remove the grass, if necessary, from your pond site. Since our pond would only cover a small area, Bear just dug up the grass with a shovel. In the background of the pond area, you can see where we already built a small berm of turf that we dug up from elsewhere in the yard. It's covered in black plastic weighted down with rocks and logs to keep weeds from growing.

2. After removing the turf, we covered the entire area with black plastic to keep the weeds away until we were ready to start constructing the pond. If you have a larger area to dig up and enough time, you could also use the black plastic method to kill the grass. If you put down plastic in the Fall you'll be ready to dig come Spring. (You can also use a sheet mulch of cardboard & straw, see our website for more details.)

2. Create a Pond Design:

To Create the Most Habitat:

- **Incorporate more variety.** The more variety you have in your pond, the more habitat it will create.
- **More edge creates variety.** An oval shape will have the least edge. An irregular shape with curves and inlets will have more varied spaces for different habitat to develop, and will look more natural.

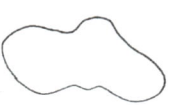
Possible pond shapes I drew.

- **Include different depths** for more habitat.
- I've read that a pond without a pump and filter **needs to be at least 43 square feet in volume** to create a balanced environment, so shoot for that if you have the space.

Include a Shallow End/Beach:

- There should be a shallow, open end without large rocks or vegetation, **where you create a beach area for wildlife to access the pond**.
- The beach **will be used for drinking** by a wide variety of animals and **birds will use it for bathing**.
- The shallow end **should be about 4-6" deep**, and have a gradually sloping beach that will allow frogs and other wildlife to enter the water.
- **Add large, flat stones** near the beach **for birds to sit and prune themselves** after bathing.

Include a Deep End:

- The deep end will provide wildlife **places to hide from the predators** that are also drawn to ponds.
- If it's deep enough to be below your frost line, it will **give wildlife a place to hibernate & help plants survive Winter**.

Include Shelves Around the Edges:

- Most ponds have underwater shelves along the sides **where plants that aren't happy in deep water can grow**.
- Shelves may be used by some frogs for mating.
- Recommendations for the size of side shelves vary, from **8-12" deep**, and from **9-12" wide**.

Include a Berm if You Want:

- If you're making a larger pond, you'll be digging up a lot of soil. Decide what you'll be doing with it. You can pile it around the pond on one or more sides to create a berm or small hill. **The berm can be planted, creating more habitat and a visually pleasing backdrop for your pond**. Just make sure it doesn't block your view, and try not to make it so steep that it looks unnatural.
- The soil **can also be used to direct any overflow of water** that might occur during heavy rains. Bear was quite worried that the pond would overflow and make our basement wet, so I built a trench to direct overflow into one of our rain gardens, which I thought was a great use of resources. However, even during one of our wettest months on record, although the pond was very full, there was never any noticeable water streaming away from it.

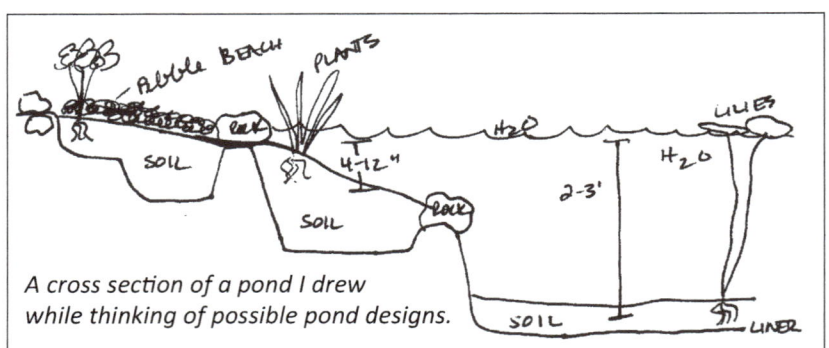
A cross section of a pond I drew while thinking of possible pond designs.

How Deep Should the Pond Be?

- Most sources seem to agree that you should aim for making your pond **at least 2 feet deep and ideally 3 feet deep to benefit the most wildlife**.

- But you'll also want to **consider where the frost line is in your area**, which is the depth at which the moisture present in the soil is expected to freeze during the colder months. **Some pond animals and plants need to be below the frost line in order to survive the Winter.**

- It seems to be difficult, though, to find definitive information about where the frost line is. I emailed my county extension office (which I found by searching online for "[County name] extension office [state name]"), but they were not able to answer my question, and were very discouraging about how steep the sides of my pond were likely to be given my pond dimensions.

- Other information I found online said that the deep end of the pond should be at least **2' if you're in USDA Gardening Zone 5** (where I am in southern Wisconsin), **or at least 3' for the colder Zone 4**.

- I should probably have tried to contact the building inspector for our village, who would know the building code requirements for how deep you need to build house foundations so they don't suffer from frost heave in the Winter (when the ground freezing makes the foundation move).

- In the end it didn't matter. To build a deeper pond, you need to increase the width of the pond so that the sides aren't too steep. We had only a small area in which to build, so we just built the pond as deep as we could. I got to 2 feet.

How Steep Can the Sides of the Pond Be?

Generally, the sides **should slope as gently as possible to insure that they won't collapse**. Specifically, the information I've found said this:

- Pond sides should be **20° from vertical**.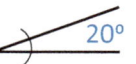
- The sides should **drop at a rate of 1/3 of the distance traveled across the top**.

I ended up building almost vertical sides in the deep end of our pond, however, in order to get more depth, because I wanted frogs to be able to hibernate there. **Our soil is clay**, though, which gives the walls pretty good strength on their own. Seven years later they still haven't collapsed, but I do worry about it sometimes, so I don't recommend doing this unless you have to.

Mark Your Pond Design on the Ground:

Use something to mark the shape of your pond on the ground. I used landscape paint from a spray bottle that I bought when I was less environmentally knowledgeable. You can use flour, corn meal, rope, or a hose if you have the more flexible types.

I wanted more curves in our pond, so it would look more natural and have more habitat. I drew a lot of really complicated, curvy shapes on paper when planning my pond, but **when I was actually marking it on the ground, I realized we had too small of an area to include lots of curves.** Also, more curves can make putting in the pond liner more tricky (you'll have to do more folding). So I ended up with the simple, but at least not oval, shape above.

3. Rough Dig the Pond

- When you've finalized your design and marked it on the ground, start digging.
- Try to **set aside the top soil** (in buckets, on a tarp, or just make a separate pile) so you can put it back around the edge of the pond at the end of the project, as it is the best soil for growing plants. You can usually tell the top soil from the deeper soil because it has a different color and/or texture.
- In the photo below, you can see in the left side of the hole where I got below the browner top soil and hit the redder clay beneath. You can use the non-top soil dirt to create a berm, which I've done to the right of the hole.

If You Run into Tree Roots:

- **Use a pruning saw** (#4 right) **to cut them a couple inches beneath the surface** so they hopefully won't poke through the pond liner later. You can reach back into the soil with the blade of the pruning saw to cut the root, you don't have to dig back that far.
- **Make sure it's a root and not a buried utility line** (see *Choose & Prepare Your Location* above).

How I Tried to Make Planting Shelves in the Pond:

- Our cat Frost helped with the digging. Below, he's laying in a "planting pocket" I made. A second pocket is across from him at the top of the photo.
- **This is how I tried to put side shelves in my pond. I wanted to put my plants directly into the soil rather than leave them in pots, so I tried digging two "pockets" or indentations on the sides, into which I put the liner and covered it with soil.**
- I did plant arrowhead in these pockets the first year, but the pockets were so small, the arrowhead jumped ship pretty quickly. I think this idea might work, though, if you could build a bigger pond that would accommodate larger pockets.

My "Planting Pockets" idea didn't work very well, although our cat enjoyed napping in them.

Useful Digging Tools

1. Regular Shovel: This is used for the initial rough digging of the pond.

2. Straight-End Shovel: If you have one of these, it can be used to begin smoothing out the bottom and sides of the pond after the rough digging is done.

3. Japanese Hand Hoe: This was the best tool for the final smoothing of the pond sides. It has a small, triangular head and a very sharp edge, which allowed me to scrape the dirt along the sides of the pond and get it very smooth (you can find one by googling the term).

4. Small Pruning Saw: This is used to cut roots. You can reach back into the soil with the blade of the pruning saw to cut the root a few inches beneath the surface, so it won't poke back out through the pond liner.

4. Level the Edges

The edges of the pond must be as level as possible.

- If they're not, the liner will stick out above the water where it will be exposed to UV rays from the Sun, causing it to degrade faster.
- Bare liner also doesn't look as nice.

Making the pond exactly level is much easier said than done. **To see how close you are:**

- **Place a carpenter's level on top of a piece of wood,** such as a 2x4, **that's long enough to go across the pond** (below). Adjust as needed by adding or removing soil.
- We needed a short piece of wood to go across the narrow width of the pond and a longer piece of wood to test the entire length of the pond.
- I've also **put a yardstick** (it's yellow, below) **against the level in order to measure pond depth.**

In the end, I apparently did not manage to get the pond perfectly level. Some of the liner shows in the deep end, but it gets shaded by plants. The shallow end loses water more than I want it too, but that may be a leak near the top (see *How to Patch a Hole in the Pond Liner*). Do the best you can.

Set a carpenter's level on a piece of wood to see if the sides are even.

Use a yardstick to measure depth.

5. Smooth the Bottom & Sides

After you have the basic hole dug:

- **remove anything sharp**, such as rocks
- **smooth out the sides and bottom of the pond** as much as possible.
- I **used a Japanese hand hoe** (see *Useful Tools* on the previous page) to scrape the sides of the pond, after which I removed by hand the soil that fell to the bottom.

You want the soil to be smooth so that nothing sharp pierces a hole in your pond liner when the weight of the water holds it down.

Below you can see the smoothed shallow end at the top of the photo as compared to the still rough end at the bottom.

SHALLOW END

Here are a couple shots of the pond after I smoothed it out by scraping all the edges and removing the loose dirt. You can see to the right that **the walls in the deep end are much steeper than they're supposed to be**, but it worked, presumably because there's so much clay in our soil.

The smoothing process reminded me of the times in my life when I've sculpted, and this part of the project was really enjoyable. **It was like making a large Earth sculpture.**

6. Install Padding to Protect the Liner

To further protect the liner from punctures, such as when you need to walk in the pond to do maintenance, you should also install padding under it. You can buy padding, called underlayment, from pond shops that sell pond liner. Alternately, you can also use newspapers, old carpeting, and/or sand.

We used newspaper, because we were able to get unsold papers in quantity from Bear's place of employment. Here's how it works:

- Try to get a **1/2" thick layer of newspaper**.
- It's best to **unfold the newspaper to its entire length**. This is a big job, so I recommend doing it before hand. Bear did a lot of it for me while he was watching TV.
- He **measured the papers into 1/2" thick stacks and stored each stack criss-crossed** so I could easily tell that I was putting it on thick enough.

- As you're installing it, **wet down the paper to shape it and keep it from blowing away**. It will make mud puddles.
- If it's hot out and your pond is big, keep hosing down the newspaper as you work, or **cover the finished part with plastic to keep it wet** (above).

I found installing the padding to be another fun part of the pond project, even though it was a bit exhausting, because it was like a big paper mache project.

7. Buy the Liner

I waited to buy my liner until I had the pond dug, since I wasn't sure before hand what size it would end up being. The liner size is determined by the length, width, and depth of your pond.

To Determine the Size of Liner You Need:
- **Length:** Take the longest length of your pond and add 4' (which gives you 2' extra on each end)
- **Width:** Take the widest width and add 4' (which gives you 2' extra on each side)
- **Depth:** Double the depth of the pond (which accounts for going up and down both sides) and add the depth to *both* the length and the width.

Example for an 8' long by 5' wide by 3' deep pond:
- Length (add 4' to the longest side): 8+4=**12**
- Width (add 4' to the shortest side): 5+4=**9**
- Double the depth (3x2=6) and add it to *both* the length and width: **12**+6=18 and **9**+6=15
- So for a 8'x5'x3' pond you need a 18'x15' liner.

Our pond was 7 feet long, 4 feet wide, and 2 feet deep, so we needed a 15'x12' liner, or 180 sq. ft.

What Type of Liner to Buy:

You can buy rigid plastic forms to create a pond, but I was interested in using flexible liners because they allow more variability in shape. Below are the types of flexible plastic liner I read about and/or came across. Some of the books I read were old, so I'm not sure if all of these actually still exist, or if there are better ones currently available:

- **EPDM rubber:** This is the type I ended up buying from a specialty pond shop. It came with a 20 year warranty, and is UV resistant & non-toxic to fish. It's very thick, so is harder to tear or puncture. It cost about $1 a square foot when I bought it in 2012.
- **Butyl liners:** These are supposed to last the longest (50 years) and cost about $2 a foot. I didn't find this for sale anywhere in our area.
- **PVC liners:** These are considered to be reasonably tear resistant and some are guaranteed for 10 years, but they're not supposed to be exposed to sunlight or they degrade.
- **Pond liner at home improvement stores:** I found something sold as "pond liner" in the garden section of our big box hardware store. It came in packages of specific sizes. It didn't say what type of plastic is was and it seemed very thin. However, if you don't have a lot of money, it's probably the cheapest you can find. It just may not last as long. The salespeople also tried to sell me roofing rubber and said it was the same thing, but I wondered if it was fish safe? On the other hand, what do they do to the EPDM rubber to make it safe for fish?

Shopping for the liner after reading about the above types was somewhat challenging. Even one garden center that was well-known for installing ponds in our area couldn't tell me what type of plastic they used for their pond liner. They seemed to expect me to just trust it was safe because they used it.

Our Pond Cost Under $300

- We bought our liner from the Paradise Pond Shop in Cottage Grove, WI (*ParadisePondShop.com*) which was the only pond shop I found that sold retail & could tell me that they had EPDM rubber.
- They sold liner on a roll, so I had to buy it in that width. I ended up with 240 sq. ft. at $.86/foot, which cost $218.00.
- Our only other expenses were a few bags of river rocks and pea gravel, a truck rental ($25) to haul the edging rocks (that we got for free from a friend), and the cost of the plants.

8. Install the Liner

- **Store the liner in the shade**, until you're ready to put it on the pond, as UV rays can damage it.

- When you're ready to install the liner, it helps to **lay it in the Sun for a little while first** (but not on any grass or plants that you don't want to kill). The Sun will warm up the liner, **making it more pliable and easier to work with**.

- The liner for our pond was **extremely heavy**, so you may need help carrying it to the pond. I couldn't have lifted it myself, but Bear was able to.

- To get your liner onto your pond, **lay it out flat and role it up like a tortilla**, which Bear is starting to do, above right.

- **Carry it to one end of the pond and unroll it** the same as if you were putting a pie crust in a pie pan.

- **Push the liner down to the bottom of the deep end of the pond.**

- Begin molding it to the sides and **fold it into creases to make it fit**— like gathering fabric to sew pleats.

- **Weigh down the folds and the bottom and edges to keep the liner in place** while you work (I used rocks). You could also put a couple inches of water in the bottom of the pond to hold the liner in place.
- **Don't worry much about how all the folds will look.** When you add water, everything will get flattened into place and the folds will quickly get covered in natural pond "muck" and you won't see them.
- This is not an easy part of the project. **Do the best that you can.**

9. Fill the Pond with Soil, Rocks, & Water

Determining Your Pond's Water Volume:

- **Before you fill your pond with water, if you can, note down your water meter reading.**
- **Then note the reading after you've filled it with water.** The difference will tell you how many gallons of water your pond holds, which can help you determine how many plants and animals your pond can support. I made a mistake when I was doing this, and recorded the reading on our old water meter that no longer worked, completely forgetting that our village had recently installed new ones.
- If your pond is less irregular in shape than ours, **you can also determine water volume this way:** one cubic foot of water is 7.5 gallons, so the pond length x width x depth x 7.5 = the pond volume.

Pond Water:

- **Rain water is best** to use for filling your pond, if you have a lot of rain barrels or you get lucky and build the pond right before a big storm.
- Otherwise, you can **use municipal water, which is likely chlorinated, so it should sit for several days** in the pond to let the chlorine evaporate before adding any plants or animals.

Pond Soil:

- I read that **clay soil** is the best to use in a pond because it stays soft in water. That's what we have in our yard, so that's what we used.
- **6" of soil on top of the liner** is supposed to be enough to grow plants.

What We Did in Our Yard:

- We shoveled around 6" of clay soil into our deep end (using some of the dirt we had dug up) then filled the deep end with water from the hose. The pond will look muddy at this point, but will clear up when the dirt settles.

- I used **river rock** (medium sized, oval stones with no sharp edges) in the narrow middle part of the pond, and **pea gravel** (tiny rounded stones) in the shallow beach area. **I used larger stones to create walls, placing them where they kept the pea gravel & river rock from sliding into the deeper areas.**
- River rocks & pea gravel can be bought in bags from the hardware store. Just make sure you **don't use any sharp edged rocks** that could pierce the liner.
- I included lots of rocks because it creates more niches for habitat and I read that having more rock surfaces helps keep down algae.

- **Keep filling the pond** with water. Begin removing the rocks from the outer edges of the liner.

- **Let the liner settle** for a bit before cutting off the excess around the edges.

- **Trim the excess liner, leaving about a foot around the edges.** You'll be able to tell if the pond is level by whether there is any liner showing above the water (ours looks level, doesn't it?). It is difficult, but to some extent, you can remove dirt from under the liner to level it out. Then cover the liner with dirt. It's pretty impossible to make a totally smooth transition between the liner and the dirt if you have steep sides like we do. You can see the liner showing a bit in the photo below. This can be shaded by plants, or covered by letting edging rocks stick out over the water a couple inches.

10. Building a Privacy Trellis Behind the Pond

We built our pond in our narrow side yard, right next to our neighbor's driveway and back deck. To improve the view, we installed a privacy trellis as a backdrop for our pond:

- Above, Bear is leveling metal 4"x4" post footings while pounding them into the ground.
- Below, we have slipped four white PVC posts into the metal footings and attached them with screws.

- Between the posts we attached brown plastic lattice. Now we can grow vines on this trellis for privacy.

Trellis Cost: We got lucky one day and found the 4"x4" white posts on sale at Habitat Restore for $10 a piece (the Restores are Habitat for Humanity's architectural salvage shops). The metal footings were around $5 a piece. The plastic lattice we also had to buy new- at around $20 a piece. So it was $120 total.

Trellis Design: Our posts were 6' tall and our lattice was 8', so **I cut the lattice into triangular shapes at the top**, which was easy to do with a jigsaw. I didn't want to lose the extra 2 feet, since our neighbor's driveway was so much higher than our side yard. However, this probably wasn't the best plan, as **the triangles started falling over when vines started growing on them**. We had to reinforce the lattice by adding the plastic edging that is sold to go with it. Later, we also had to anchor the tops of the triangles by running wire between them and the cedar T-posts we put up behind the pond on which we have tried to grow a hardy kiwi.

11. Adding a Frog Condo

I wanted to **create shelter for frogs & toads near the pond**. They like **shady, cool places** to get out of the hot Sun, and some hibernate by digging into the soil. I created this frog condo to try and meet these needs, based on suggestions I found in wildlife habitat books:

- Place some **old clay pots half-buried on their sides**.
- Next, **position rocks on and around the pots to make chambers** between the pots where the frogs could also hide and/or dig deeper.
- Then completely **cover the condo with rocks, leaving some tunnel entrances** near the front.

Unfortunately, I put the condo near the deep end of the pond. If I were doing things over, I would **put it near the shallow end**, as frogs can exit the pond most easily from there. If you want to build something more simple, **a nearby mound of twigs & leaves will also provide shelter**.

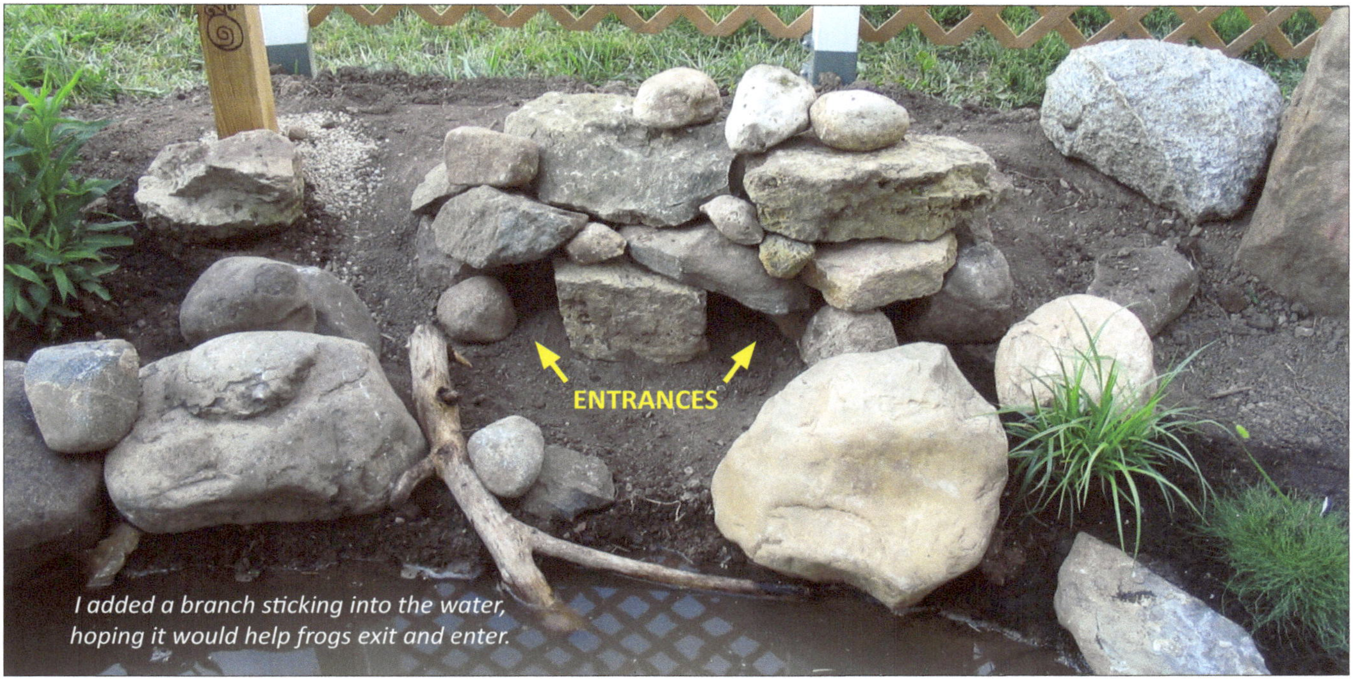

ENTRANCES

I added a branch sticking into the water, hoping it would help frogs exit and enter.

12. Adding Rocks Around the Edge of the Pond

The liner at the edge of the pond should be covered either by rocks or soil & plants to protect it from the Sun. If you're adding rocks, which we chose because we love the way they look, consider the following:

- **Leave spaces** on the pond edge **where you can kneel to do pond maintenance.**
- Stones with texture won't be as slippery when wet.
- Some stones can be **placed so they extend over the pond water by 2"** so the liner can't be seen.
- When laying the stones, **create as many small crevices as you can** in which creatures can shelter.
- Try to arrange the stones to **make shady spots** for frogs to get out of the Sun.

Transporting the Rocks:

- We don't own a vehicle that could carry heavy rocks, so we **rented a truck** from Home Depot. In 2012, it cost $20 for an hour and a half rental.
- **We had a bit of an adventure with the truck.** They come with an alarm that goes off when you've overloaded the truck (it held 2000 pounds). We had no idea how many rocks would be too much, but it didn't go off while we were loading the rocks or during the 30 mile drive home. However, when we pulled into our driveway, the alarm started blaring. It was very loud, like a car alarm, and too loud to stand next to. It didn't stop even when Bear turned off the truck. We couldn't figure out what to do. Bear ran inside and called the store while I tried to frantically unload rocks, but since they were heavy, it was slow going. It was such a relief when the alarm stopped! Home Depot gave us a discount on the rental for our troubles, so it turned out to be a small blessing.

- We got the above lovely landscaping rocks from a friend, who had a pile of them in her yard that she wanted to get rid of. We just had to dig them up. This was not an easy task, but it was sure worth the effort to be able to use them around the pond.

- The truck came with a loading ramp that has ridges across it. To get the really large and incredibly heavy stones onto the truck, we levered them out of the ground with a crow bar and positioned a dolly underneath them. Using the wheeled dolly, Bear pulled the rocks up the ramp, while I tried to push. We needed to lean on the ridges part way up the ramp and rest, because they were too heavy to pull all the way up at once, even for Bear.

Before:

Is Bear so tired (below) from moving rocks, that he's napping on them? No, he's actually installing our largest rock as a standing stone behind the pond.

After:

When arranging rocks, remember to leave as many nooks and crannies as you can to provide habitat for small creatures.

I love arranging rocks. Each one has its own beauty. I seem to have an instinct about where each one should go, although sometimes I do have to rearrange them a few times until it feels right. Take your time listening to your rocks and maybe they will tell you where to put them, too.

13. Our Finished Pond, June 2012

A view from the other direction showing the trellis:

Additional Ways to Build Wildlife Habitat

To improve the habitat value of your pond, do as much as you can of the following:

Help Animals Safely Access Your Pond:

- **Don't put large rocks around the entire edge of the pond**, it makes it too hard for small animals to get into or out of the water. One book even said that frogs can stick to hot rocks! I designed my pond with two open areas around the edge, one with pea gravel and the other with low growing plants. In a later year, one of our big rocks fell into the pond and we left it there to create a new beach area, which was used almost immediately by the larger birds, who stood on the rock to take baths.
- **Place large sticks at a couple different spots on the edge of the pond in the deep end.** Leave one end of the stick on the ground out of the water and anchor it with rocks so it won't move. Let the other end trail into the water at a gentle slope. **This can help any wildlife (including insects) that are stuck in the deep end of the pond get themselves out.**
- **Put in some vertical sticks for dragonflies to land on.** I stuck a few bamboo garden stakes into a pot full of dirt that I put in the bottom of the pond so that I don't poke through the pond liner (I don't remove the pot in the Winter). The bamboo sticks almost straight out of the water, and as you can

see (left), the dragonflies do like them.

- You can **add flat rocks just under the water surface** for birds to stand on while bathing.

Create Shelter:

- **Arrange the rocks in and around your pond to leave small niches and tunnels**. This will provide shelter, as well as access and escape routes.
- **Build a mound of twigs and leaves outside the pond** but near the edge so toads will have a place they can hide and burrow into.
- **Allow a moderate amount of dead leaves to accumulate in the pond** to be used as breeding and hiding places for amphibians and other creatures.
- Frogs need dense, shady ground vegetation for shelter after they leave the pond, so **plant sedges, rushes, or native grasses around the edges.**
- I've read that you should put a pile of big sticks in the deep end of the pond to provide places for frogs to hide from animals that will eat them. This makes sense, but the pieces of wood I tried to add just floated, and I didn't want to wedge anything into the mud at the bottom and risk piercing the liner.

Other Suggestions:

- **Muddy areas at the edge of the pond will be used by various wildlife.** Butterflies gather at sheltered mud puddles (called puddling) to drink minerals from the mud. Muddy areas also benefit birds & bees (such as robins & non-aggressive mud dauber wasps) that use mud for nest building. Pond edges will usually stay muddy, although sometimes the mud falls into the pond and you need to add soil.
- **Plant night blooming flowers** to encourage moths, which will provide food for frogs.
- **Inoculate your pond:** I went to a nearby natural pond on public property and got a gallon of water from it. I added the water to my pond in order to "inoculate" my pond water with whatever normal microscopic organisms there are in ponds.

Above is a close up of the pebble beach end of our pond. The shallow water and gentle slope lets this frog and other creatures easily get into and out of the pond. The underwater flat rocks on either side are used by the birds when drinking.

Below you can see one of the logs stuck into the deep end of the pond to help birds & other creatures access the water. It can also help bugs or small animals get out of the pond if they're stuck. I may need to pile rocks onto the back end of the log to keep it in place.

Above: Birds love to bathe in the pebble beach and we love watching them enjoy it.

Below: This rock jutting out into the shallow end is balanced on a wide piece of PVC and some other rocks, so that there is a tunnel underneath it that can shelter animals in the water. Animals above the water sit on the rock and drink. Eventually the rock broke along the crack you can see in this photo, but then the end that fell into the water made a perfect ramp for birds and animals to get to the water and drink.

Wildlife Seen at Our Pond

First Two Weeks: Once we finished building our pond at the end of May (2012), I was incredibly excited to see wildlife begin using it right away:

- Birds used it for drinking & bathing.
- In less than a week there were little beetles living under the water. Where did they come from?
- Mud dabber wasps, who are not aggressive, came to take mud. They use it to build tubes in which they lay their eggs.
- I saw dragonflies laying eggs in the water.

Two Months Later:

By July I was amazed to see a lot more life in the pond:

- There were bigger underwater beetles and dozens of tiny snails.
- I found dragonfly shells stuck to cattail stems.

Early Frog Sightings:

- By early August I saw our first frog! I wasn't expecting them to move in until the following Spring. It was a Green Frog. The next night I saw it sitting in the shallow end, probably waiting to catch bugs.
- In 2014 we heard a peeper frog in the Spring and a couple tree frogs calling during the Summer.
- By 2015, we had a much larger chorus of frogs serenading us in the Spring, and we saw tadpoles in the pond for the first time!
- Subsequent years have all varied. Sometimes we hear several frogs, sometimes just one. Sometimes we have tadpoles, sometimes just adult frogs.

The first frog in our pond, 2012.

Our first (Eastern American Toad) tadpoles, 2015.

Below is some of the information we've learned about our most frequent visitors, dragonflies and frogs.

Dragonflies

- **Dragonflies have been around for a long time.** There are fossils resembling them that are 300 million years old, making them one of the planet's earliest insects.

Bear took this shot of a dragonfly we saw on a tour of a small, urban permaculture farm in Milwaukee that included a backyard pond.

- **Dragonflies are very good hunters.** They have excellent eye sight (with almost 360 degree vision) and a sharp bite. Their long legs, which aren't very good for walking, help them hold insects they've captured in flight.
- **Their wings are very powerful,** allowing some of them to fly at speeds up to 60 mph. Each of their four wings is able to move independently, allowing them to fly in any direction, including sideways and backwards, to catch a meal.

- Adult dragonflies have very short lives (weeks or months).
- You may also see **damselflies** visit your pond. They are smaller than dragonflies, with thinner bodies. When at rest, they hold their wings back, in line with their bodies, whereas dragonflies hold them out at a 90º angle away from their bodies. Together the dragonflies and damselflies make up the Order Odonata.

I didn't realize when I took this photo of a beautiful insect that I was looking at a damselfly.

Dragonfly Lifecycle

- **Dragonflies mate in the air, then lay their eggs in still water.** If you see a dragonfly hovering over your pond and repeatedly dipping the end of her tail into the water, you're watching her lay eggs.
- **The eggs turn into larvae (called nymphs or naiads) that live underwater,** voraciously eating mosquito larvae, tadpoles, and even small fish.
- Some larvae live under water for two or more years, before becoming adults, but we saw some metamorphizing within a couple months after we built our pond.

Empty dragonfly nymph shells.

- **The nymphs crawl out of the water onto plants** or the shore, **where they break out of their shells, spread their new wings, and fly away.** The photo to the left shows some empty nymph shells clinging to the miniature cattails in our pond.

A Dragonfly Rescue

- In July of 2014, I was fortunate enough to go outside just after a dragonfly had emerged from its shell, which is shown on the stem of the arrowhead plant below. The shell was still green at this point, because it was so new, but it will soon turn brown.
- Unfortunately, the dragonfly that emerged had somehow fallen into the pond. It was fluttering its wings, but was unable to break away from the water.

- I lifted it out with our small rake (below) and was able to enjoy seeing it up close for a few minutes (after running in the house shouting for Bear to come see) before it flew away.

The dragonfly below has just emerged from its shell, above.

OurTinyHomestead.com • 27

Frogs

- **Frogs are amphibians, which means they need two habitats to survive: both water and land.**
- They are **born in the water, as tadpoles, where they are fully aquatic**. They breathe underwater and have no limbs, just a tail that helps them swim.
- When the tadpoles **undergo metamorphosis**, they begin growing legs. Eventually, they lose their tales and emerge from the water. They **begin breathing air, and start their life on land**. Since you built them a backyard pond, you get a front row view to this miracle.

  - If you're very lucky, you'll be around when the new frogs come out of the pond. It can be like a migration, with dozens of frogs emerging at once. They are tiny (less than 1/2", above), and for a while you have to be careful not to step on them as you move around the garden.
- Where do all those baby frogs go? I don't know the answer to that, but I hope a lot of them stick around, because my yard provides them a lot of habitat, but my neighbors' yards are mostly lawn.
- **Adults must be near water.** Even if they live mostly on land, they need water to breed. Males find an appropriate body of water and call to attract a mate. When they breed, the male clasps the female while fertilizing the eggs she releases.
- Adults can breathe through their semi-permeable skin, so also need to stay near water to keep their skins moist. Such thin skin absorbs toxins (e.g., pesticides & herbicides), which is why we hear a lot about frogs that are found with deformities. They are an early warning sign that the land is in peril.
- Adult frogs are carnivorous. They eat many insects, so are appreciated by gardeners everywhere.
- **Frogs are "cold-blooded" (ectothermic)**, which is somewhat of a misnomer. It's not that their blood is cold. Instead, they acquire their body heat from outside sources. They must move into the Sun when they need to warm up, or into the shade when they need to cool down, so make sure the landscape around your pond provides them with both options.
- Frogs have a flat round spot behind each eye, their **tympanic membrane, which is used for hearing.**
- In Wisconsin, where I live, we have 12 types of frogs. So far, we've been able to identify the following three in our yard:

Cope's Gray or Gray Treefrog

- These two treefrogs are hard to tell apart, so we're not sure which one Bear found sheltering under his grill cover (right).
- **ID:** They are most easily distinguished by their bright yellow inner thighs & large toe pads that help them cling to vertical surfaces. They can be gray, brown, or green, changing color based on the temperature or their surroundings.
- **Adult Size:** 1.25"–2"
- **Call:** This is the best way to tell them apart. The Gray Treefrog call is like a high pitched trill of a bird. Cope's is a quicker, tighter trill, that is more insect like, and not as pretty (search online to hear the difference). They both call on warm, humid days, not just when they are breeding.
- **Habitat:** Adult treefrogs dwell in trees or bushes, returning to water to breed.
- **Fun Fact:** They produce an "antifreeze" that allows them to survive Winter, then thaw out in Spring.

Northern Green Frog

- **ID:** They are most easily recognized by the green on their upper lips and the prominent dorsolateral fold that runs down their bodies. Males have yellow chins and their tympanum (ears) are smaller than their eyes, whereas in females it is larger. They are common throughout Wisconsin.

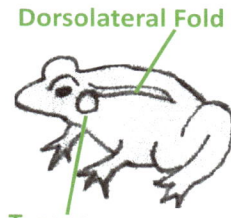

Dorsolateral Fold
Tympanum

- **Adult Size:** 2.4"–3.5"
- **Call:** is made from the end of May through early August. It sounds like a rubber band/banjo string being plucked, and is usually given as a single note.

- **Eggs:** are laid in a mass attached to vegetation at the water's surface.
- **Habitat:** They spend most of their time in the water, even overwintering there. They prefer water with heavy vegetation.
- **Tadpoles:** Green frog tadpoles do not metamorphize until their second Summer. They overwinter under water. The blurry photo to the left shows them around our pond heater in the middle of the Winter (see *Winter Pond Maintenance*).

Eastern American Toad

- Toads are actually a type of frog. Their skin is thicker than other frogs, allowing them to live farther from the water, where they can eat 32,000 garden insects in a season.
- **ID:** These toads are common throughout Wisconsin. They have chunky bodies and bumpy skin (that doesn't cause warts). They are best identified by the large, protruding paratoid gland behind each eye that is used in self-defense. When a toad is picked up by a predator, the glands secrete a fluid that causes a burning sensation in predator's mouth, encouraging them to quickly drop the toad.

Paratoid Gland

- **Adult Size:** 2"–3.5"
- **Call:** is a distinctive long, continuous trill, made from late April through early July.
- **Eggs:** They are the only frog in Wisconsin to lay their eggs in long strings.

- **Tadpoles:** are solid black, making them easy to distinguish. They are the only tadpoles in Wisconsin that form schools. They metamorphize within weeks.

Pond Plants

I have tried growing many different types of plants in our pond, that I have bought from both garden centers and online stores. I grow them directly in the soil at the bottom of the pond, because I am trying to recreate a natural pond, and many have not survived the Winter (we are in USDA Gardening Zone 5, which gets down to -20° F/-29 C°). However, I kept trying new plants until I found ones that survived, because **plants are a necessary component of the pond ecosystem:**

- They shade the water and help **keep it from growing excessive amounts of algae**.
- Plants also **oxygenate the water**, which helps keep it from getting stagnate and smelly.
- Plants on the edge of the pond **provide cover** and shade for amphibians.

This is the hardest thing to digest about pond plants:

- **Most of the pond surface (2/3 to 3/4) should be covered in plants.**
- So, after you build a wonderful water feature, you need to hide most of the water. The shade provided by the plants is an important component in preventing algae blooms, especially in ponds without pumps (see the *Pond Maintenance* section).

Types of Pond Plants:

The next thing to understand about pond plants is that there are several types, each playing a different role. The most balanced pond will have most of these types, although some plants fulfill multiple functions:

- **Filtrators:** these plants absorb nutrients so that algae doesn't grow in excessive amounts. Some filtrators I've tried growing include: water hyacinth (but it can be invasive), watercress (but it prefers running water in streams), cattails (which have edible & useful parts but spread a lot), and arrowhead (*Sagittaria* species, which has edible tubers).

- **Submerged:** the leaves of these plants are under the water, where they breathe in carbon dioxide and breathe out oxygen. These oxygenators are some of the most important types of plants to have in your pond. They help keep the water from becoming stagnant and smelly.

The above submerged plant (that floats under the water) was not labeled when I bought it from the garden center, but it has persisted for years in our pond.

- **Floating:** these float on top of the water and help prevent algae blooms by shading the water. Their roots dangle in the water and help absorb excess nitrogen and phosphates. Most floaters are annuals (except duckweed never seems to die).

Above is duckweed (the tiny plant covering the surface that you'll see in almost any pond) and water hyacinth (the larger, bulbous plant). Both plants float on the surface of the pond and trail their roots into the water below, helping to keep it clean. Water hyacinth (Eichhornia) is prolific- two plants thrown into the water in the Spring covered the surface of our small pond by Summer. Don't let them escape into the wild, where they can choke waterways. They did not survive our Winters.

- **Water-Lily Like Plants:** these grow in the soil under the water, but their leaves float on the surface (water lilies) or rise above it (lotuses and arrowhead). Their large leaves shade the water and can provide landing pads for wildlife. I particularly like lotuses. The water beads on their leaves beautifully and they have some edible parts, but I could not get them to overwinter in our pond (they probably need deeper than our 2 feet of water).
- **Marginals:** these grow at the edges of ponds, in either moist soil or shallow water. If planting them in soil, consider whether you have enough soil over the pond liner to support the plants you want to use. You may need to use small ground covers with shorter roots right next to the pond, which also makes it easier for you to see the water. Grasses, sedges, or rushes planted farther back can break up the look of shorter plants & provide wildlife cover.

Above: We've planted thyme at the edge of the pond, which we hope will become a low ground cover, letting us see the water better on the side of the pond where we walk the most often. The terra cotta pot is meant to provide shade for frogs in the meantime.

Below: these are the plants growing in & around our pond that have survived our Winters (2019).

Because I follow permaculture principles, I always try to grow plants that are useful to me as well as to wildlife. I found sections on edible pond plants in:

- *Perennial Vegetables: From Artichoke to "Zuiki" Taro, a Gardener's Guide to Over 100 Delicious, Easy-to-Grow Edibles,* by Eric Toensmeier. ISBN: 978-1-931498-40-1. It includes a section on "The Edible Water Garden" and in-depth descriptions of many of the plants.
- *Toolbox for Sustainable City Living* by Scott Kellog & Stacy Pettigrew. ISBN: 978-0-89608-780-4. It has a section on plant aquaculture, but is also worth reading for the rest of the content.

Food From Our Pond: Arrowhead Tubers

- Arrowhead (*Sagittaria latifolia*) is a submerged water plant that is easy to grow & has **very tasty tubers.**
- It is named after its leaves, which are triangular shaped, like an arrow.

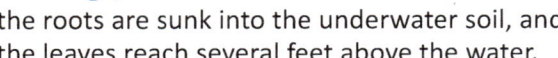

- **It likes to grow in full Sun, at the water's edge**, where the roots are sunk into the underwater soil, and the leaves reach several feet above the water.
- It's a colony forming root crop that most years fills our whole pond.

- Arrowhead is a perennial native that grows throughout the U.S. Its tuber was eaten by Native Americans.

How to plant Arrowhead:

- You can push the tuber into the underwater soil in the Spring, with the growing tip (the hook-like part that is growing out of the bulb) facing up.
- If you're planting something that has leaves, you'll need to put some small rocks around the plant, to keep it from floating away, until the roots grow enough to support the plants.
- **It's all right if all the leaves are completely submerged during planting.** You'll be amazed at how quickly they grow to reach the surface.

- I received mail-order tubers in the Winter, and they didn't come with any growing instructions. I decided to plant them in pots with potting soil. Then I sat the pots in a shallow pan that I kept filled with water, and put the hole thing under a grow light in our small greenhouse. They survived, so I transplanted them to the pond in early Spring.
- It might also work to store the tubers in the refrigerator in a bag with some water in it if you can't plant them right away.

2015: the arrowhead grew so much you couldn't tell there was a pond underneath.

Overwintering:

- To overwinter, the **tubers should not freeze solid,** so your pond needs to either be deeper than the frost line in your area, or you need to use a pond heater. The first time I planted arrowhead it didn't survive the Winter, but I tried again another year, when I had a better pond heater, and I was very happy to discover a couple arrowhead tubers with tiny leaves floating in the pond the next Spring. There always seem to be a few that become dislodged for some reason, every year (left).

I had trouble with this plant floating to the surface but replanted it & weighed it down with rocks.

Harvesting Arrowheads:

1. Harvest the plants in the Fall, when they have died back. The photo below is from Oct. 11, 2015. You can see that all the arrowhead leaves are gone (they usually fill the pond), and all that remains are the brown stems, which you can see on the left, leaning against the rocks.

2. According to what I've read, Native American women harvested arrowhead by walking into the frigid water and using their bare feet to dig around in the mud for the tubers. *Brrr.* We get to cheat, since our pond is so small. We can lay on the edge and dig around with just our hands, although it's still really cold.

3. Unfortunately, the tubers don't grow at the end of the dying stems, so you can't find them by just pulling up the plant stems, like Bear is trying below.

4. You have to dig around in the muck until you find the tubers.

5. Bear loves a good treasure hunt, and has longer arms, so he dug around the deep end.

6. If you pull out any little tubers, it's best to replant them so they'll be enough to regrow next year.

7. Here's what the tubers look like. The growing tips from which young plants will emerge reach out from the round bulb.

> **TIP:**
> Don't wait too long after the plants have died back to harvest them. We've found if we do, the flavor changes for the worst, which happened with the tubers below. The skin around them was darker than usual, so maybe it started to go bad?

Roasting Arrowhead "French Fries"

1. To clean the tubers you harvested, begin by rinsing them in a bowl of water.

2. Next, pull off some of the loose brown (and sometimes gooey) skin from the tuber and the growing tip.

3. Use a brush to remove the reddish brown layer that is on the outside of some parts of the tuber. You can get most of it off with a gentle scrub.

4. These tubers are cleaned and ready to cook.

5. Trim off the bottom (root end) of the tuber & the growing tip.

6. A look at the insides, which are rather rigid but with very small air pockets... I hesitate to say it's like styrofoam, but I can't think of any better analogy.

7. Chop the tubers into small squares and put them in a baking pan. We also added some optional garlic cloves & ground pepper.

8. Coat with a little olive oil and bake at 425° F for 45 minutes.

9. The arrowhead will brown a bit on the outside and get crunchy. The inside is light and airy, and they are like eating a french fry that isn't oily.

10. We ate ours arrowhead "fries" as the side dish for some burgers. They were great dipped in our homemade ketchup. We liked them so much we began planning another pond so we could grow more.

11. Uncooked tubers can be stored in the refrigerator in a bag with a little water to keep them moist.

Arrowhead Growing Challenges:

- Rabbits like to eat the leaves, so, initially, don't plant them too close to the edge of your pond.
- Birds may use the stems to land on and inch down so they can get a drink of water. This may bend the stems. It doesn't seem to hurt the plant, but it does look a bit messy (below left).
- The stems sometimes get infested with tiny insects, although this doesn't seem to affect growth either.

Other Reasons to Grow Arrowhead:

- Dragonfly nymphs use the stems to crawl out the water (below middle) when they go through metamorphosis.
- They have pretty white flowers (below).

Natural Pond Maintenance

People seem to worry a lot about what their pond water looks like (evidenced by all the chemicals you can buy to clean it). When I first put in the pond, I worried too (mostly that I wouldn't be able to keep it clean without using those chemicals that are potentially harmful to wildlife). However, by building as natural a habitat as possible, and employing some patience while letting Nature do Her work, I have fairly easily maintained a healthy pond (without commercial products). Here are some of the ways I've managed that:

Avoiding Stagnant Water:

- Pond pumps are popular because they keep the water from becoming stagnant. Water flowing from a pump down a waterfall mixes oxygen into the water as it splashes, keeping the water fresh so it doesn't become smelly.
- Since wildlife ponds don't have pumps, you instead **aerate the pond with underwater plants** that put oxygen into the water (see the *Pond Plants* section).
- Alternately, you could also **add a small water dripper, sprayer, or fountain to the shallow end.** If you make sure the water it uses goes back into the pond, it will help aerate it. The sound of the water will also help more wildlife find your pond. A sprayer will be used by birds (including humming-

I am trying a round, floating solar powered fountain in my pond to attract more wildlife.

Our pond with a normal Spring algae bloom.

birds) to take baths. They are sold as accessories for bird baths or ponds. I am experimenting with the inexpensive solar powered versions that are now available. So far, I don't think the pump they use is strong enough to injure any tadpoles.
- When you have to add water to your pond, **make sure the water from your hose or bucket splashes into the pond, which also adds oxygen.**

Dealing with Algae:

Some ponds get algae blooms. This happen when there is so much algae growing in the pond that the water is completely full of it and it's floating on top. People seem to find this unsightly and feel they need to do something about it (by adding chemicals), but there are some times when algae blooms are normal, and you should just wait them out:

- **Algae blooms are normal in a new pond** before the system balances itself. This can take a couple months, so be patient.
- **Algae blooms are normal each Spring**, before the plants have grown back after dying off each Winter. Just wait it out and the pond will soon return to normal after the plants start growing.

A balanced pond does not stay clogged with algae. The plants and animals and other parts of the ecosystem keep it under control. Here's what can help balance your pond:

- **Warm water helps algae grow, so keep the majority of the pond's surface area covered with plants.** This keeps the Sun from warming the water.
- **Snails eat algae:** I tried to buy some snails when we first built our pond (they're sold in pet stores for aquariums and online for ponds), but no one could tell me if the snails would survive our Winters. It turns out I need not have bothered, though, snails moved into our pond very quickly on their own, like magic. Where did they come from?
- **Tadpoles eat algae:** Our pond is the most algae free in the years when we have the most tadpoles.
- **Organisms in the soil** you've put in the bottom of your pond are supposed to **help keep the water clean** too.

What Doesn't Work Well for Algae:

- You can **manually remove algae** that's floating in the pond. This can be done with a toilet brush, or we use one of those tiny leaf rakes that are about 8" across (below), that are meant to get between bushes. The algae you pull out is good fertilizer, and we spread it around the garden. However, you **have to be careful— you might also remove and so kill tadpoles, dragonfly nymphs, and snails that get caught up in the algae.** If you leave the algae at the edge of the pond for a while, some creatures might be able to crawl back in, but it's best to try to remove them with your fingers before spreading the algae around.

- **Don't use barley straw:** Bags of barley straw meant to float in your pond are advertised as a natural way to control algae. I tried this, but our pond seemed to get worse. Then I noticed that the fine print on barley bag said that it **doesn't work in stagnant water**, where it will start killing the good bacteria. I removed the barley and our pond improved quickly.

When the Pond Water is Low:

One of the reasons I wanted a pond was that I was tired of filling the bird bath every day. Unfortunately, pond water evaporates, so if it doesn't rain often, I still have to add water. There are things to consider when doing this:

- Remember that if you have **municipal water**, it **may have a high chlorine content and you don't want to add a lot of it to your pond at once** or it could hurt the plants, animals, and microbes.
- My books said that you shouldn't add more than a tenth of the volume of the pond or several inches of chlorinated water at once. Instead, they recommend only adding a bit of chlorinated water each day, or letting the water sit in buckets for one or two days so that the chlorine evaporates. This all seemed a bit cumbersome to me, so in the beginning I just added water often so it never got very low, and didn't notice any problems.
- I later discovered that you can **buy a filter for your outside hose that removes the chlorine** (search online for "water filter for garden hose"), and have since been using that.
- You could also **add water from a rain barrel**, so you don't have to worry about the chlorine content.
- As stated above, when adding water, don't put the hose under the water. **Place the hose so that the water gently splashes onto the surface** (above), which **helps keep the water aerated**.

Controlling Mosquitoes:

- **Mosquitoes can only breed in still water.** This is one of the big reasons most people want a pump in their pond and why in areas with mosquito spread diseases, there may be laws against having still water ponds. However, in a balanced pond, there will be wildlife that are happy to remove the mosquito larvae by eating them, so a pump is not actually required for mosquito control.
- In the first few weeks after we built our pond, I noticed mosquito larvae in the shallow end. Larvae begin as hundreds of tiny brown dots suspended in the water. As they get larger, they hang upside down from the surface of the water and you can see them squirming around. I had read that if you squirt the surface of the water with the hose once or twice a week, it will disrupt their development cycle and the larvae will die. I tried this and it seemed to work. We are blessed with very few mosquitoes in our yard (thanks to bats & frequent breezes), so if hundreds of mosquitoes had hatched from our pond, we likely would have noticed.
- **Dragonflies also only lay eggs in still water.** Shortly after I attempted mosquito control with the hose, I saw a dragonfly dipping its tail in our pond, which meant it was laying eggs. The dragonfly nymphs that hatch from the eggs and live underwater eat mosquito larvae, and thereafter we didn't see any more mosquito larvae (see the *Dragonflies* section).
- When frogs move into your pond, they will also eat mosquitoes (both the tadpoles and adults).
- I've read that minnows eat mosquitoes. I discussed this with a person from a local bait shop, and he thought that the minnows would not live long in a small, warm pond without running water. And I didn't want them to die over Winter, so I did not try adding minnows to our pond.
- Another suggestion I read is that caffeine kills mosquito larvae. It said to put plain powdered coffee in the pond, just enough to change the color of the water slightly. I haven't tried this, so don't know if it works, or if it affects other life in the pond.
- When you're thinking about mosquito control, keep in mind that they provide a rich and needed food source for the wildlife that will use your pond, so your goal should be to keep a balanced system where all types of life are present, but none become harmfully overabundant.

Getting Into the Pond:

- Walking into the pond to perform maintenance can be daunting. The bottom is slippery & full of rocks, and it's such a different world that it's hard not to find it a bit scary wondering what is living down there.
- In Spring of 2014 one of the large rocks on the edge of our pond fell into the water, and Bear was the brave one who waded into the deep end to lift it out.

- But it occurred to me that if we rested it on top of the planting pocket we'd built in the pond, it would make a great step, so Bear obligingly put it back in. We now have a more convenient entrance and another shallow area that allows more wildlife to access the pond.

Achieving a Balanced Pond

After building our pond, I struggled to help it find a balanced state, which to me meant using the natural methods described here to create a pond in which:

- the plants overwinter without my having to bring them indoors and replant them every Spring
- the water stays clean, smells good, and doesn't have abnormal amounts of algae
- the wildlife takes care of any mosquitoes breeding in the pond.

What Was Our Pond Missing?

- **The key to balancing our pond turned out to be finding the right plants** (see the *Pond Plants* section), since Nature obligingly sent the right animals soon after the pond was finished.
- It took a while for me to find plants that survived the deep freezes of our Winters. I tried many traditional pond plants: lily pads, lotus, cattails, and others I don't even remember, but none came back after the Spring thaw. Eventually, I planted arrowhead (*Sagittaria latifolia*), and that same year found a better Winter heater for the pond (see the *Winter Maintenance* section), and the plant has come back every year since.

Arrowhead

- The other plants I was missing were the type that float in the water. They cost $3.95 each at the garden center, and you're supposed to put one per every 2 square feet of surface area in your pond. I figured that I would need 5 or 6 and that seemed too expensive too have to replant every year. In 2015, though, the pond seemed to have too much algae, so I finally broke down and bought some. The plants weren't even labeled at the garden center so I didn't know what I was buying (and still don't). They looked like seaweed:

- After I added the floaters to the water that Spring, it eventually became as clear as it has ever has been. I could see tiny tadpoles swimming in the water all the way from the inside the house.
- When I went to put the unnamed floaters in the pond, I realized that each bunch was held together with a rubber band, and actually held lots of strands of whatever they were. So I really only needed one bunch. By mid-Summer, they had taken over the pond, and I had to start trying to remove some, which was hard to do without removing other life from the pond.
- As it turned out, the floaters weren't annuals, and they have grown back in the pond each Spring since.

It took three years for our pond to become balanced, after which it became generally self-sustaining.

Winter Pond Maintenance: Keep Part of the Surface Unfrozen

In the Winter, I keep part of the pond surface unfrozen so the birds and mammals can still access drinking water. But even if you don't want to provide water for wildlife, **you shouldn't let ice completely seal the surface of your pond. It can cause any wildlife hibernating under the water to suffocate.** At least a small vent hole is needed to let gases escape. Also, it may be coincidence, but my pond plants didn't start overwintering until I used the right pond heater.

I read about several ways to keep the pond from icing over, but I think they must have been written by people living in warmer climates. They would not work here in southern Wisconsin, but maybe they will work where you live:

- Float something in the pond like a log or a ball.
- Put a twig in the pond and jiggle it every day.
- Set a hot pot of water on the pond and let it melt through the ice every day.

Choosing a Pond Heater:

In our cold climate (USDA Gardening Zone 5, which gets down to -20° F / -29° C), none of the above methods are sufficient, so I use an electric heater instead. I have tried several types over the years:

- The first Winter I used a small **bird bath heater** that was 200 watts. It **worked to keep the shallow end of the pond open**, but I wasn't sure if there was any way for the gases from the deep end to escape, where the frogs, if any, would be hibernating.

- The second year I bought a **floating stock tank heater (1500 watts)**, which is used for keeping large tanks of drinking water unfrozen for farm animals. I put it in the deep end, and it **was strong enough

Small hole in the pond ice that was hard to keep open, made by a non-floating heater.

My boot

to keep the entire pond unfrozen**. However, **our electrical bill went up $50 a month and the water was warm enough to evaporate** so I had to keep carrying heavy buckets from inside the house to replenish the water.

- Next, I bought a **non-floating, 300 watt pond de-icer**. It was meant to work for up to a 300 gallon pond, but since it didn't float, it **had trouble keeping an opening near the surface**. It made a small hole in the ice (above), but only because I poured boiling water in it every day.

- Currently we're trying a **1250 watt "Floating Pond De-Icer"** that seems to be doing a better job of keeping only part of the deep end of the pond unfrozen. We got it from our local farm store.

What I've learned through trial & error:

- Use a heater that floats.
- Put the heater in the deep end of the pond, not the shallow end.
- Experiment with what wattage works best. The lower the wattage, the more money and electricity you save, and only a small vent hole is needed to keep wildlife safe.
- If you keep too much of the pond unfrozen, the water will start evaporating and will need to be replenished.

How We Safely Plug in Our Pond Heater:

The heaters I've bought have all come with short cords, usually about 15 feet long, which wasn't anywhere near long enough to reach from the pond to our nearest outside electrical outlet, but the instructions always say using it with an extension cord would lead to unspecified dire consequences.

Since we normally have snow on the ground in Winter, I wasn't very worried about the cords starting a fire, but I learned that we still needed to be careful. One year, when I tried to remove a thin extension cord connected to the heater, I found that the ends of the two cords had melted together. Not good!

So Bear asked an electrician if he could extend a heater cord for us. Instead, he said that as long as we **use one of the thicker (very expensive) extension cords (15 amp, 125 volt)** and that we **keep it very, very dry**, it would be safe to use and not burn out the heater. So here's what has worked for us:

1. Attach a heavy duty extension cord to the pond heater's cord:

2. Wrap the entire connection tightly with black electrical tape (which can be purchased at hardware stores) to keep out water:

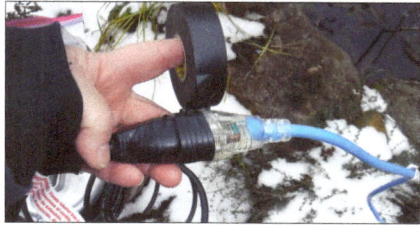

3. Put the taped cords into a sealable plastic bag:

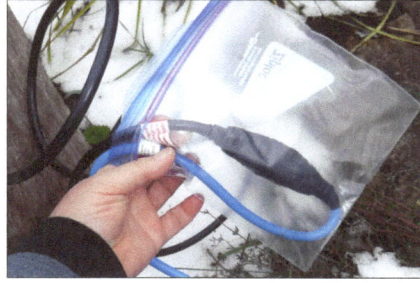

4. Seal the bag as much as you can, fold it around the cords, and wrap it closed with duct tape:

5. Hang the plastic bag above the ground using a shepard's hook, so it won't get wet lying in any melting snow:

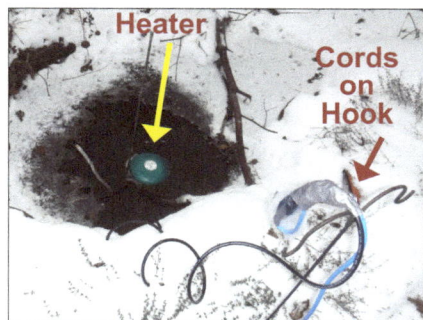

6. I also drape the rest of the cord over hooks so it's above any snow between the pond and the house's electrical outlet:

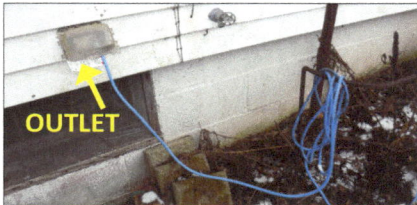

Please get advice from your own electrician if needed, or, if you can afford it, have them install an outlet next to your pond, which is the safest and easiest alternative.

OurTinyHomestead.com • 41

How to Patch A Hole in the Pond Liner

One year I began to notice that after I filled the pond with the hose it would lose a couple inches of water overnight. So I began to suspect that there was a hole in the liner, and eventually found one a couple inches across. I have no idea how it got there. (I read that one person had holes because a deer stepped on his liner, but I don't think that's our problem.) I did a really poor job trying to patch it the first year I found it, but on my second attempt a couple years later, I succeeded.

Day 1: Clean & Flatten Liner

1. Read the instructions for your pond patching kit, which you can buy online. Mine included a bottle of primer, a red-handled wooden roller, a green scrubbing pad, blue rubber gloves, and 2 black patches.

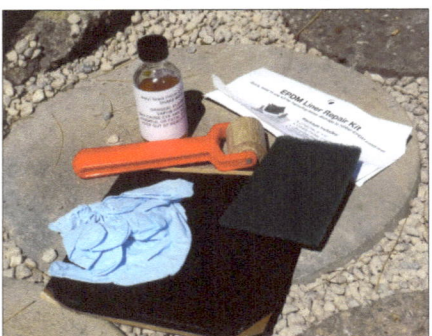

2. Expose the liner around the hole. This is the hardest part- you may need to remove surrounding soil, plants, rocks, or water.

3. The liner must be flat- I put a brick under it to create a flat surface that the patch would stick to.

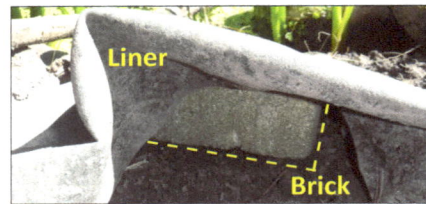

4. Scrub thoroughly around the hole with the pad. It must be clean so the patch will bond to it.

5. I put a brick over the liner to further flatten it, but didn't cover the hole because it would get moist underneath while sitting overnight.

Day 2: Apply Patch

6. Use scissors to cut your patch so that it is 2" wider than your hole.

7. Remove the backing from the patch. It will be very sticky! So be careful how you touch it and where you lay it down.

8. The instructions said to shake the primer well. I then put some on the scrubbing pad and wiped it around the hole. I was supposed to wait until it was tacky before applying the patch. In the hot, direct Sun, it took me a while to realize this happened almost instantly.

9. Apply the patch and press it down. Use the roller that should come with your kit to make sure it's smooth. Roll over the entire patch, paying special attention to the edges.

10. I then put another brick over the top of the patch to keep it flat while the patch & liner bonded. I hoped the patch wouldn't also stick to the brick below the hole (see *Patching Tips*).

11. Let dry completely before exposing to water. The instructions didn't say how long this would be, so I made sure it wasn't supposed to rain and didn't add water for a couple days, but leaving it alone overnight was probably sufficient.

PATCHING TIPS:

- When I tried to remove the brick from underneath the patched area the next day I realized the patch had bonded to it. I was able to get it unstuck, but this problem could have been easily avoided if I had put a piece of paper on top of the brick & under the hole in the liner before I applied the patch.
- **Try not to apply the patch in the direct Sun.** The heat made the primer dry almost instantly and it made the edges of the patch pieces melt and stick to the plastic bag they came in.
- **Don't apply a patch over a fold.** Initially, I did not remove enough landscaping to be able to make the liner flat (right). I actually tried applying a patch over the fold, but of course, it came off within days. For my second attempt, we were removing the landscaping for other reasons, so it was easier to pull out all the liner surrounding the hole & flatten it.
- **Use primer.** I originally bought a cheaper patch kit without primer, but the primer seems to makes a more secure bond.
- **Check the weather to make sure it's not going to rain** for at least a few days after you apply the patch.

The yellow star shows where the hole was patched.

OurTinyHomestead.com • 43

The Pond in Later Years & Through the Seasons

If you're a gardener, you'll know that your pond is going to look quite different as the seasons change and the years go by. I prefer low maintenance gardening that mimics Nature, which means that sometimes the plants overgrow their space or the weeds flourish during the height of Summer. At these times, I remind myself that **although the pond may not look the best to us humans, it never stops providing good wildlife habitat**. Your pond can always look more manicured if that's what you prefer.

Spring 2013

July 2013

May 2014

July 2015

Fall 2018

March 2014

Summer 2019, when we put in a new path next to the pond, giving the area a more manicured look that some prefer.

Conclusion

It is my hope that this book has given you the tools and encouragement you need to build a pond, so you can start enjoying wildlife in your own backyard.

I believe that if we could all be close to Nature on a daily basis, rather than spending 95% of our time in buildings and cars, we would begin to remember that we are part of the web of life on this planet. We would stop feeling separate from the rest of Nature, and we would start undoing the damage we are causing to other species. In my yard, and in my life, I try to follow my favorite quote, which is from Gandhi:

Be the Change You Wish to See in the World

We can all begin making a difference in our own backyards!

~ Theresa Berrie
2019

To Learn More:

We would love to hear about your experiences. Contact us through our website, where you'll also find more ideas about gardening for wildlife:

OurTinyHomestead.com

Also consider these programs:

Certify your yard as wildlife habitat with the **National Wildlife Federation**
www.nwf.org

Pledge to provide a humane backyard for wildlife with the
Humane Society of the United States
www.humanesociety.org

Our Tiny Homestead

Theresa & Bear are trying to homestead as much as possible on a tiny lot (60' x 140') in a tiny town in southwestern Wisconsin, where they bought a fixer-upper in 2005.

They are searching for a way to live that connects them to the seasons and honors & protects the Earth. Theresa mostly contributes the inspiration & brains, while Bear mostly brings the brawn and the much needed humor. They share their adventures at OurTinyHomestead.com

Printed in the USA
CPSIA information can be obtained
at www.ICGtesting.com
LVHW072006110224
771576LV00002B/14